ANOVA by Example
Hands on Approach using SAS

Faye Anderson, MS, PhD

Copyright © 2019 Author Name

All rights reserved.

ISBN: 9781710316513

DEDICATION

To my family.

CONTENTS
1 INTRODUCTION   1
2 WHAT IS ANOVA   7
3 ONE-WAY ANOVA   11
4 TWO-WAY ANOVA   20
5 K-WAY ANOVA   23
ABOUT THE AUTHOR   28

1 INTRODUCTION

Since analysis of variance is a type of hypothesis testing, it is imperative to have an overview of hypothesis testing.

In studies where we are comparing groups, we statistically test the difference between them by conducting hypothesis testing. An example would be the difference between groups of patients: one took a placebo, another took drug A, and another took drug B.

In hypothesis testing, there are two hypotheses: the null hypothesis (H0) and the alternative hypothesis (Ha). In this context, H0 states that the population parameter (e.g. the mean or variance) is equal for the groups. Whereas the null hypothesis states that the groups' parameters are not equal.

SAS for Free
SAS offers free license at the following link:

https://www.sas.com/en_us/software/university-edition/download-software.html

p-value Interpretation

Every hypothesis test produces a probability value or p-value which assesses how compatible the sample data are with the null hypothesis. High p-values indicate that the data is compatible with H0, whereas low p-values indicate that the data is not compatible with H0 (or agrees with Ha).

Significance Level (alpha)

In order to understand how to interpret p-value, it is important to get acquainted with the concept of significance level (alpha or critical value). This is the probability of rejecting the null hypothesis when it is true, or the probability of committing a type I error. With a significance level of 0.05, there is a 5% chance of rejecting a true null hypothesis. If the significance level is higher than the conventional 0.05, such as 0.10, this will increase the chance of a false positive to 0.10, but it will also decrease the chance of a false. If the significance level is lower than the conventional 0.05, such as 0.01, this decreases the chance of a false positive.

Type I and Type II Errors

The following table demonstrates the four scenarios between having true (or not true) H0 and making the correct or incorrect decision. Alpha is the probability of committing type I error or rejecting a true H0 whereas beta is the probability of committing type II error or failing to reject a false H0. The statistical power of the test is equal to 1 - beta.

	H0 True	H0 False
Decision 1: Fail to reject H0	Correct Decision	Incorrect Decision = Type II Error (Beta)
Decision 2: Reject H0	Incorrect Decision = Type I Error (Alpha)	Correct Decision

The following table summarizes how to interpret p-value in general.

Scenario	Decision	Conclusion
p-value <= significance level (0.05)	Reject the null hypothesis (H0).	Ha is true
p-value > significance level (0.05)	Fail to reject the null hypothesis (H0).	H0 is true.

Commonly used Hypothesis Tests

This section demonstrates the following important four tests.

Test	Purpose
T-test	Compare the means of two populations
F-test	Compare the variances of two populations
Kolmogorov-Smirnov	Tests whether the data is normally distributed or not, when the sample size is large or when there are no outliers in the sample.
Shapiro-Wilk	Tests whether the data is normally distributed or not, when the sample size is relatively small (n < 50) or when there are outliers in the sample.
Levene's	Tests whether the groups' variances are equal or not

T-test in SAS

The following code demonstrates how to use proc ttest in SAS to conduct t-test. The output shows the p-value of <.0001, which is less than the default significance level of 0.05. So, we reject the null hypothesis H0. In this example, H0 states that the mean of the variable year-rev is equal to zero. We conclude that the mean of year_rev is not 0.

One-Sample t-test Example:

```
/* t test */
proc ttest data=WORK.IMPORT1 sides=2 h0=0 plots=none;
    var year_rev;
run;
```

Output:

Variable: year_rev

N	Mean	Std Dev	Std Err	Minimum	Maximum
51	65.8765	6.7344	0.9430	37.4000	75.9000

Mean	95% CL Mean		Std Dev	95% CL Std Dev	
65.8765	63.9824	67.7706	6.7344	5.6347	8.3714

DF	t Value	Pr > |t|
50	69.86	<.0001

The following example demonstrates how proc ttest can be used to compare the means of two variables/groups ozone and wind. T-test assumes the variances of the two groups are equal. Since p-value is less than 0.05, we reject the null hypothesis that the two means for ozone and wind are equal.

Proc univariate was used to test for normality. The small p-values (less than 0.05) force us to reject the null hypothesis of normality for the difference between ozone and wind.

Two-Sample t-test Example:

```
data Work._Paired_diffs_;
      set BOOK.AIRQUALITY;
      _Difference_=Ozone - Wind;
      label _Difference_="Difference: Ozone - Wind";
run;

/* Test for normality */
proc univariate data=Work._Paired_diffs_ normal mu0=0;
      ods select TestsForNormality;
```

```
        var _Difference_;
run;
```

Output:

Variable: _Difference_ (Difference: Ozone - Wind)

Tests for Normality				
Test		Statistic	p Value	
Shapiro-Wilk	W	0.889226	Pr < W	<0.0001
Kolmogorov-Smirnov	D	0.153452	Pr > D	<0.0100
Cramer-von Mises	W-Sq	1.061138	Pr > W-Sq	<0.0050
Anderson-Darling	A-Sq	5.810486	Pr > A-Sq	<0.0050

Difference: Ozone - Wind

N	Mean	Std Dev	Std Err	Minimum	Maximum
153	29.2320	31.0456	2.5099	-12.4000	130.9

Mean	95% CL Mean		Std Dev	95% CL Std Dev	
29.2320	24.2733	34.1908	31.0456	27.9134	34.9758

DF	t Value	Pr > |t|
152	11.65	<.0001

2 WHAT IS ANOVA?

Analysis of variance also known as ANOVA is a type of statistical hypothesis testing that compares two or more groups' means. It has types and different uses as will be explained in the coming sections.

Types of ANOVA

There are three types of ANOVA: one-way ANOVA, two-way ANOVA, and k-way ANOVA. Some books classify ANOVA in two categories by combining the two later types into one group. The table below demonstrates the differences between these types.

ANOVA Type	Uses	Example
One-way ANOVA	When we need to compare only one factor or independent variable between the groups.	If we want to compare whether the mean pay of three workers is the same based on their working hours.
Two-way ANOVA	When we need to compare two factors between the groups.	If we want to compare whether the mean pay of three workers is the same based on two factors: their working hours and years of education.
K-way ANOVA	When we need to compare k (more than 2) factors between the groups.	If we want to compare whether the mean pay of three workers is the same based on few factors: their working hours, gender, number of previous jobs, and years of education.

Assumptions of ANOVA

ANOVA is a parametric test. This means that its assumptions need to be verified prior to running the analysis. The following are the three ANOVA assumptions.

Assumption 1

The observations are independent. This means that the data or the sample was collected randomly. It is important to understand that there is no way to test for independence of observations in SAS or using any statistical software. This assumption can only be enforced by correctly randomizing the sampling method or experimental design.

Assumption 2
The observations are normally distributed. This can be tested using the Kolmogorov-Smirnov or the Shapiro-Wilk tests.

Assumption 3
Homogeneity, which is the equality of the population variances for the groups. This can be tested using Levene's test for the homogeneity between groups.

ANOVA vs t-tests
Both methods compare means. However, ANOVA tests for means equality for more than two groups/samples. If there are three groups (A, B, C) and we wanted to test if their means are equal, ANOVA will do the comparison effectively without worrying about maximizing type I error. Here is how: If we conduct a t-test in this example then we will need to run it three times, for groups A and B, A and C, and for B and C. The more hypothesis tests we use, the more we risk making a type I error, and the less power a test has.

Limitations of ANOVA
The result of an ANOVA analysis is that there is a significant difference between groups, not which groups are significantly different from each other. This can be accomplished by conducting a post-hoc comparison

to find out where the differences are significantly different from each other and which are not. Example post-hoc comparisons are Scheffe's and Tukey's.

3 ONE-WAY ANOVA

One-way ANOVA compares the variance in the group means within a sample when the question is about one independent variable or factor. It compares three or more levels of a categorical factor to decide whether there is a difference between their means.

Air Quality Dataset

This dataset contains daily air quality measurements in New York, May to September 1973. It can be downloaded from R:

https://www.rdocumentation.org/packages/datasets/versions/3.6.1/topics/airquality

The dataset has four continuous variables for the measurements of ozone, solar radiation, wind, and temperature. It has two categorical variables: month and day. The data was collected for 5 months, from May through September. Following are the summary statistics.

Analysis Variable : Ozone Ozone				
Mean	Std Dev	N	Skewness	Kurtosis
39.1895425	29.3333275	153	1.1754125	0.8058269

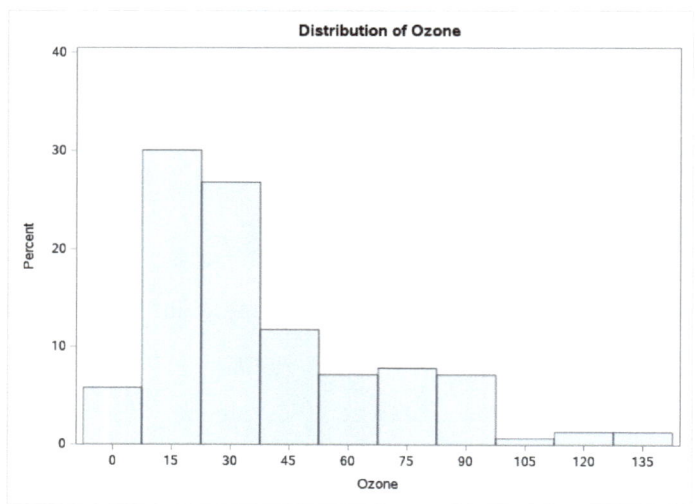

Before we conduct one-way ANOVA to test if ozone levels were equal in the five months, we need to verify the three assumptions of randomness, independence, and homogeneity.

Randomness is assumed in this documented study. It is obvious that ozone is not normally distributed because of the skewness of its histogram. Moreover, its skewness is far from zero and its kurtosis is far from 3. Thus, we transform it using the natural logarithm function before conducting three types of ANOVA: using PROC GLM, using PROC ANOVA, and using PROC npar1way.

The null hypothesis for this procedure states that the monthly means for ozone are equal from May through September. The alternative hypothesis states that at least one month's mean is different from the remaining four.

Transforming data Example:

```
data work.transform;
        set WORK.AIRQUALITY;
        log_Ozone=log(Ozone);
run;
```

One-way ANOVA Example using Proc GLM:

```
ods noproctitle;
ods graphics / imagemap=on;

proc glm data=WORK.TRANSFORM plots=none;
        class Month;
        model log_Ozone=Month;
        means Month / hovtest=levene plots=none;
        lsmeans Month / plots=none;
        run;
quit;
```

Output:

Class Level Information		
Class	Levels	Values

ANOVA by Example

Class Level Information		
Class	Levels	Values
Month	5	5 6 7 8 9
Number of Observations Read		153
Number of Observations Used		153

Dependent Variable: log_Ozone

Source	DF	Sum of Squares	Mean Square	F Value	Pr > F
Model	4	18.0280650	4.5070163	7.94	<.0001
Error	148	84.0373784	0.5678201		
Corrected Total	152	102.0654434			

R-Square	Coeff Var	Root MSE	log_Ozone Mean
0.176632	22.32070	0.753538	3.375962

Source	DF	Type I SS	Mean Square	F Value	Pr > F
Month	4	18.02806501	4.50701625	7.94	<.0001

Source	DF	Type III SS	Mean Square	F Value	Pr > F
Month	4	18.02806501	4.50701625	7.94	<.0001

Levene's Test for Homogeneity of log_Ozone Variance ANOVA of Squared Deviations from Group Means					
Source	DF	Sum of Squares	Mean Square	F Value	Pr > F
Month	4	2.2654	0.5663	0.70	0.5921
Error	148	119.5	0.8072		

Level of Month	N	log_Ozone	
		Mean	Std Dev
5	31	2.80184426	0.89379840
6	30	3.44318004	0.76204484
7	31	3.79463204	0.70023652
8	31	3.60494963	0.72770282
9	30	3.23275442	0.66004115

Least Squares Means

Month	log_Ozone LSMEAN
5	2.80184426
6	3.44318004
7	3.79463204
8	3.60494963
9	3.23275442

One-way ANOVA Example using Proc ANOVA:

PROC ANOVA DATA=WORK.TRANSFORM;
CLASS Month;
MODEL log_Ozone = Month;
TITLE 'Compare Ozone across Months';
RUN;;

Output:

Compare Ozone across Months

Class Level Information		
Class	Levels	Values
Month	5	5 6 7 8 9

Number of Observations Read	153
Number of Observations Used	153

Compare Ozone across Months
Dependent Variable: log_Ozone

Source	DF	Sum of Squares	Mean Square	F Value	Pr > F
Model	4	18.0280650	4.5070163	7.94	<.0001
Error	148	84.0373784	0.5678201		
Corrected Total	152	102.0654434			

R-Square	Coeff Var	Root MSE	log_Ozone Mean
0.176632	22.32070	0.753538	3.375962

Source	DF	Anova SS	Mean Square	F Value	Pr > F
Month	4	18.02806501	4.50701625	7.94	<.0001

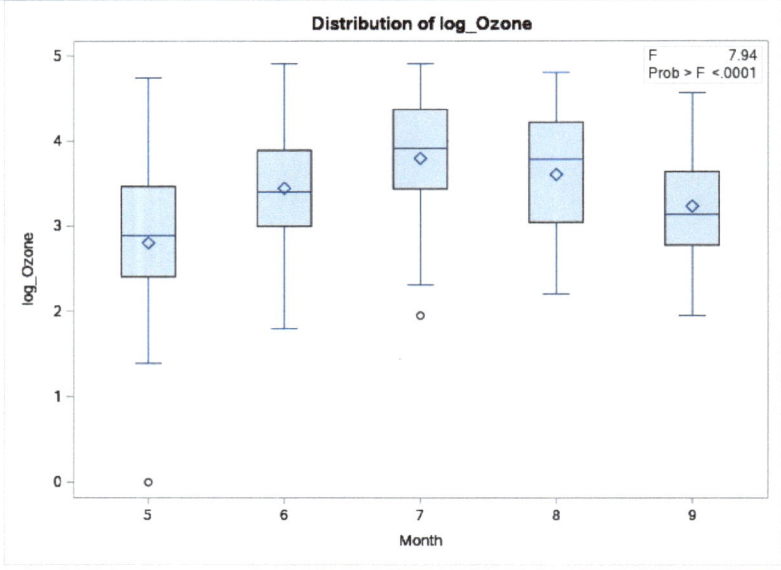

Non-parametric One-Way ANOVA Example:

ods noproctitle;
proc npar1way data=WORK.TRANSFORM wilcoxon
plots(only)=(wilcoxonboxplot);
    class Month;
    var log_Ozone;
run;

Output:

Wilcoxon Scores (Rank Sums) for Variable log_Ozone Classified by Variable Month					
Month	N	Sum of Scores	Expected Under H0	Std Dev Under H0	Mean Score
Average scores were used for ties.					
5	31	1515.00	2387.0	220.257547	48.870968
6	30	2396.00	2310.0	217.562089	79.866667
7	31	3127.50	2387.0	220.257547	100.887097
8	31	2754.50	2387.0	220.257547	88.854839

Wilcoxon Scores (Rank Sums) for Variable log_Ozone Classified by Variable Month					
Month	N	Sum of Scores	Expected Under H0	Std Dev Under H0	Mean Score
9	30	1988.00	2310.0	217.562089	66.266667

Interpretation of Results:

The small p-value of less than the default significance level of 0.05 allows us to reject the null hypothesis. That is, at least one month's ozone mean is different than the remaining four.

Moreover, the high p-value for the Levene's test for homogeneity tells us that the equal variances assumption is violated in this data.

Membership Dataset

The dataset Membership has two continuous variables duration and price; and two categorical variables plan and area. The table below presents the description of the variables.

Variable	Description
Membership	Number of weeks the user has been a member with the company
Area	Categorical variable for the three areas of service: A, B, and C.
Price	Amount of money the customer pays weekly.
Plan	Categorical variable for whether the

	customer is subscribed with the plan: 1 for yes and 0 for no.

The question we will attempt to answer is whether the average number of weeks of membership are the same between the three areas. In order to conduct ANOVA without worrying about the normality assumption, we will use the non-parametric procedure npar1way.

Non-parametric One-Way ANOVA Example:
ods noproctitle;

proc npar1way data=WORK.MEMBERSHIP wilcoxon plots(only)=(wilcoxonboxplot);
    class Area;
    var Duration;
run;

Output:

Wilcoxon Scores (Rank Sums) for Variable Duration Classified by Variable Area					
Area	N	Sum of Scores	Expected Under H0	Std Dev Under H0	Mean Score
Average scores were used for ties.					
A	482	515470.50	518632.0	12010.6933	1069.44087
B	1000	1100424.00	1076000.0	14366.6017	1100.42400
C	669	698581.50	719844.0	13333.7882	1044.21749

Kruskal-Wallis Test		
Chi-Square	DF	Pr > ChiSq
3.3523	2	0.1871

Interpretation of Results:

The p-value of higher than 0.05 indicates that we cannot reject H0. Hence, we conclude that the mean weeks of membership are the same for the three areas.

4 TWO-WAY ANOVA

In two-way ANOVA, there are two categorical independent variables that we suspect they impact the independent variable or the outcome.

Two-way ANOVA has three null hypotheses: that the means of observations grouped by the first factor are the same; that the means of observations grouped by the second factor are the same; and that there is no interaction between the two factors (or the interaction between the two factors has no impact on the dependent variable).

In the Membership dataset, we would like to investigate the impact of both the area and plan on duration of membership. Here we have three null hypotheses:

First H0: The means of Duration in weeks grouped by plan are the same. There are 2 means for Duration based on the two options for Plan.

Second H0: The means of Duration in weeks grouped by Area are the same. There are 3 means for Duration based on the three areas A, B, and C.

Third H0: There is no interaction between Area and Plan to impact the mean Duration in Weeks.

Two-Way ANOVA Example:
ods noproctitle;
ods graphics / imagemap=on;

proc anova data=WORK.MEMBERSHIP;
    class Plan Area;
        model Duration = Plan Area Plan*Area;
    run;

Output:

Class Level Information		
Class	Levels	Values
Plan	2	0 1
Area	3	A B C
Number of Observations Read		3333
Number of Observations Used		2151

Dependent Variable: Duration Duration

Source	DF	Sum of Squares	Mean Square	F Value	Pr > F
Model	5	5505.742	1101.148	0.68	0.6371
Error	2145	3463350.166	1614.615		
Corrected Total	2150	3468855.908			

R-Square	Coeff Var	Root MSE	Duration Mean
0.001587	39.87694	40.18228	100.7657

Source	DF	Anova SS	Mean Square	F Value	Pr > F
Plan	1	48.273626	48.273626	0.03	0.8627
Area	2	5352.331625	2676.165812	1.66	0.1909
Plan*Area	2	105.136320	52.568160	0.03	0.9680

Interpretation of Results:

The "model" statement lists Duration as dependent variable, Plan and Area as independent variable. Furthermore, Plan*Area represents the interactions between Plan and Area.

The last table of the results gives the p-values which are all greater than alpha of 0.05. That is, we fail to reject the three null hypotheses that the three factors of Plan, Area, and Plan*Area do not have a statistically significant effect on the duration of membership (weeks). In other words, the three null hypotheses are true.

5 K-WAY ANOVA

This is also known as the n-way ANOVA. This method demonstrates whether there are significant main effects of the independent variables and whether there are significant interaction effects between independent variables (factors). Interaction effects occur when the impact of one independent variable depends on the level of the second independent variable.

The number of null hypotheses for this analysis depends on the number of independent variables (factors). It is the number of the factors in addition to the interactions between them. For example, if we are testing the impact of four independent variables X1, X2, X3, and X4 on the outcome Y then we will have the following 10 null hypotheses:

First H0: The means of Y grouped by X1 are the same.

Second H0: The means of Y grouped by X2 are the same.

Third H0: The means of Y grouped by X3 are the same.

Fourth H0: The means of Y grouped by X4 are the same.

Fifth H0: There is no interaction between X1 and X2 to impact Y.

Sixth H0: There is no interaction between X1 and X3 to impact Y.

Seventh H0: There is no interaction between X1 and X4 to impact Y.

Eighth H0: There is no interaction between X2 and X3 to impact Y.

Ninth H0: There is no interaction between X2 and X4 to impact Y.

Tenth H0: There is no interaction between X3 and X4 to impact Y.

Income Dataset

This dataset has the following variables:

Variable	Description
Income	Amount of annual income in dollars.
Female	1 for female and 0 for male.
Married	1 for married and 0 otherwise.
Loan	1 for having a load and 0 otherwise.

Let us conduct an n-way ANOVA to test the impact of the three categorical variables female, married, and loan on the outcome income.

The 6 null hypotheses are as follows:

First H0: The means of Income grouped by Loan are the same.
Second H0: The means of Income grouped by Female are the same.
Third H0: The means of Income grouped by Married are the same.
Fourth H0: There is no interaction between Married and Female to impact Income.
Fifth H0: There is no interaction between Loan and Female to impact Income.
Sixth H0: There is no interaction between Loan and Married to impact Income.

N-Way ANOVA Example:
ods noproctitle;
ods graphics / imagemap=on;

proc anova data=WORK.INCOME;
 class Loan Female Married;
  model Income = Loan Female Married Loan*Female Female*Married Loan*Married;
  run;

Output:

Class Level Information		
Class	Levels	Values
Loan	2	0 1
Female	2	0 1
Married	2	0 1
Number of Observations Read		600
Number of Observations Used		600

Dependent Variable: Income Income

Source	DF	Sum of Squares	Mean Square	F Value	Pr > F
Model	6	1348800230	224800038	1.36	0.2304
Error	593	98322572101	165805349		
Corrected	599	99671372331			

Source	DF	Sum of Squares	Mean Square	F Value	Pr > F
Total					

R-Square	Coeff Var	Root MSE	Income Mean
0.013532	46.78291	12876.54	27524.03

Source	DF	Anova SS	Mean Square	F Value	Pr > F
Loan	1	662945974.5	662945974.5	4.00	0.0460
Female	1	56673625.1	56673625.1	0.34	0.5590
Married	1	7009910.6	7009910.6	0.04	0.8372
Loan*Female	1	276592340.7	276592340.7	1.67	0.1970
Female*Married	1	316202739.2	316202739.2	1.91	0.1678
Loan*Married	1	29375639.5	29375639.5	0.18	0.6740

Interpretation of Results:

The p-value for Loan is slightly less than 0.05, which suggests that we reject this null hypothesis. That is, the grouping of income data based on loan is not statistically significant.

The remaining large p-value (greater than alpha of 0.05) suggest that the remaining null hypotheses are not true. That is, the means of Income grouped by Female are the same; the means of Income grouped by Married are the same; there is no interaction between Married and Female to impact Income; there is no interaction between Loan and Female to impact Income; and there is no interaction between Loan and Married to impact Income.

These results showed us which factor is significant to income or not but did not tell us which one is of more impact on income compared to the others. This will need other analyses.

ABOUT THE AUTHOR

Faye Anderson, MS, PhD is a published author of many peer-reviewed publications and books. She has graduate degrees from Colorado State University and the University of Texas School of Public Health and have been working as a consultant and a statistician for more than twenty years.

OTHER TITLES BY THE AUTHOR

1. Anderson, F. (2016). Categorical Data Modeling by Example, Hands on approach using R, CreateSpace Independent Publishing Platform.
2. Anderson, F. (2016). GeoStatistics by Example, Hands on approach using R, CreateSpace Independent Publishing Platform.
3. Anderson, F. (2016). Hypothesis Testing by Example, Hands on approach using R, CreateSpace Independent Publishing Platform.
4. Anderson, F. (2016). Logistic and Multinomial Regressions by Example, Hands on approach using R, CreateSpace Independent Publishing Platform.
5. Anderson, F. (2016). Statistics by Example, Hands on approach using R and/or Excel, CreateSpace Independent Publishing Platform.
6. Anderson, F. (2016). Survival Analysis by Example, Hands on approach using R, CreateSpace Independent Publishing Platform.
7. Anderson, F. (2017). Biostatistics by Example, Hands on approach using R, CreateSpace Independent Publishing Platform.
8. Anderson, F. (2017). Clinical Trials Statistics by Example: Hands on approach using R, CreateSpace Independent Publishing Platform.
9. Anderson, F. (2017). Time Series Analysis by Example: Hands on approach using R, CreateSpace Independent Publishing Platform.

www.ingramcontent.com/pod-product-compliance
Lightning Source LLC
Chambersburg PA
CBHW040342220526
45473CB00009B/2768